PATRICK H. ARMSTRONG

Discovering
Geology

SHIRE PUBLICATIONS LTD

CONTENTS

ACKNOWLEDGEMENTS

Photographs are acknowledged as follows: Peter Hoare, plates 1 and 4; Aerofilms, plates 10, 11 and 12; Institute of Geological Science (Crown Copyright), plates 2, 3, 5, 13, 14 and 15. Plates 6, 7, 8 and 9 are by the author. The figures are by Michael Hadley, partly after M. Bradshaw, I. Evans, M. Smith, the British Museum (Natural History) and other sources, including the author. Cover design by Michael Hadley. Thanks are due to Peter Hoare and Edward Armstrong who read part of the manuscript.

INTRODUCTION

Why study geology?

Geology may be defined as the study of the rocks of the earth's crust. It is natural for man to be interested in his own origin and that of the world he inhabits. The rocks contain a record of the history of the planet that is man's home and through studying them at least some of our natural curiosity can be satisfied. Fossils—the remains or traces of past life—are preserved in the rocks, enabling deductions to be made about the animals and plants which inhabited the earth millions of years ago. Further study can reveal how climatic conditions have changed and how the positions of continents and oceans, mountains and deserts, glaciers and swamps have altered throughout geological time. Of special interest are the skulls and bones that have been found in many parts of the world but particularly in Africa, that tell of the evolution of man from an ape-like ancestor, and the artefacts that complete the story — man's subsequent development of tools, first of stone and later of bronze and iron.

As man's technology became more sophisticated so his requirements for raw materials and energy increased. Today geology is partly an 'applied' science; the information it provides is used in the discovery of metal ores, coal, oil and natural gas and by engineers deciding on sites for dams and power stations. Hydrogeology is the branch of the subject concerned with water supply.

In this book both 'pure' and 'applied' aspects of the subject will be considered. The origin and structure of the earth, its crust and the nature of the rocks and minerals of which it is composed will be discussed and some of the changes that have occurred since its formation will be examined and related to changes that can be seen taking place today. Emphasis is placed on examples from the British Isles in the sections on historical geology and fossils, but a few examples from outside Britain will be included for comparison.

The origin and structure of the earth

It was formerly maintained by some scientists that the earth and other planets were derived from solar material. A wandering star, it was argued, passed close to the sun and the gravitational pull so exerted caused a long filament of material to stream out into the gap between the two bodies. According to this theory, as the visiting star receded the filament broke up and the material of which it was composed condensed to form the earth and planets.

3

This view is not now widely held; the chemical composition of the sun is known to be very different from that of the earth and as the stars are so far apart the possibility of the sun and another star having come into such close proximity is considered very remote indeed.

In the last three decades astronomers have preferred a theory that the planets formed by the condensation and aggregation of particles within a swirling cloud of dust and gas that existed around the sun — a 'cool' rather than a 'hot' origin. The heat deep within the earth today is due to radioactivity and is not the residual heat of a gradually cooling planet.

It has been calculated, from a study of the amount of radioactivity in rocks, that the oldest rocks in the world (in Greenland) formed rather less than 4,000 million years ago. Rocks brought back from the moon by the Apollo missions and meteorites (fragments of material from outer space that form meteors or 'shooting stars' as they partially burn up in the earth's atmosphere) have ages of about 4,600 million years. It therefore seems reasonable to assume that the earth, moon and other bodies in the solar system formed at about the same time.

Deductions concerning the structure of the inner earth can be made from seismology, the science that deals with earthquake waves, for the different layers found within the earth transmit these waves at different velocities. The outermost layer is the crust; it is about 30-40 kilometres (about 19-25 miles) in thickness beneath the continents and about 10 kilometres (6 miles) thick beneath the oceans. The crust is itself subdivisible into two layers of differing character, the upper one of granite and granite-like rocks, the lower layer of basaltic material. Below the crust lies the mantle, the composition of which is uncertain, but it seems to consist of a much heavier material than the surface crust; a rock called dunite has been suggested. The mantle is some 2,900 kilometres (1,800 miles) in thickness and surrounds a core, perhaps of nickel-iron. The core is at least partly liquid and extends to the centre of the earth, 6,400 kilometres (4,000 miles) from the surface. The pressures in the centre of the earth are enormous — at least one million times the atmospheric pressure at the surface — and so it is impossible to speculate usefully about the properties of material found there — conditions are so different from anything at the surface.

The detailed consideration of the inner layers of the earth is the province of the geophysicist rather than the geologist, and so from this point onwards this book will concentrate on the crust.

1. THE MATERIALS OF THE EARTH'S CRUST

Minerals and rocks

Minerals are chemical substances that occur naturally in the crust of the earth. Rocks are masses of one or more minerals.

Three great classes of rocks are distinguished: igneous (those that solidified from the molten state); sedimentary (those formed as the result of the erosion of pre-existing rocks and the re-deposition of the resulting material); and metamorphic (rocks formed by the effects of pressure and/or heat on existing rocks).

Properties of minerals

(Note: a few chemical terms and formulae are included in the sections which follow. Readers not familiar with them should refer briefly to the Appendix in which they are explained.)

Many minerals may be identified in hand specimens from a few of their physical properties such as the following:

Fig. 1.   A — hexagonal crystals of quartz; B — cubic crystals of fluorite.

CRYSTAL FORM: this is important as it reflects the way in which the atoms are arranged within the molecule of the mineral substance. Crystals are bounded by flat faces and the angle between analogous faces of two crystals of the same mineral will be the same no matter what the size of the crystals or how distorted they are in other respects. Various crystal classes are distinguished, e.g. cubic, hexagonal, etc. (see figure 1).

HARDNESS: this is the resistance of the mineral to abrasion or scratching. The somewhat arbitrary but useful Mohs' scale of hardness is frequently used; ten minerals are listed in ascending order of hardness:

1. Talc (softest)	6. Orthoclase feldspar
2. Gypsum	7. Quartz
3. Calcite	8. Topaz
4. Fluorite	9. Corundum (e.g. sapphire)
5. Apatite	10. Diamond (hardest)

They may be remembered from the following sentence: 'The Great Cowardly Fox Ate Our Quaint Tiny Chickens Daintily' — the first letter of each word is the same as that of the minerals in the list above! A mineral will be scratched by any mineral with a higher number on the scale. The approximate hardness of a mineral may, however, be gauged using a finger-nail (hardness about 2.5), a copper coin (3), or piece of glass (5.5).

SPECIFIC GRAVITY (i.e. the relative weight of a mineral): the number of times heavier it is than an equal volume of water (e.g. gypsum 2.3, barytes 4.5).

COLOUR: a poor guide to the identity of a mineral as minute quantities of impurities may completely alter its appearance in this respect. Thus fluorite may appear yellow, red, green or purple depending on the impurity present. It may, however, sometimes be useful. The 'streak', or colour of the mineral in powdered form, is sometimes a better indication.

CLEAVAGE: this is the way in which some mineral crystals tend to split to form flat, smooth planes. The number and positions of cleavage planes are related to the atomic structure of the mineral and hence to the crystal form. Galena (an ore from which lead is obtained) cleaves to form small cubes.

FRACTURE: this is the breakage of a mineral in some way other than along planes of cleavage; many minerals, for example, break to reveal a curved, shell-like surface — conchoidal fracture (e.g. flint).

Rock-forming minerals

A very limited number of minerals makes up the greater part of the rocks exposed at the earth's surface.

QUARTZ: this is the commonest mineral. Chemically, it is silica — SiO_2. It is often colourless and glass-like, although it may be milky, rose-coloured, smoky or violet (in the last case it is known as amethyst). It has a hardness of 7 and a specific gravity of 2.6. Cleavage is not well-developed, a conchoidal fracture sometimes being apparent. It may form hexagonal crystals terminated by pyramids (see figure 1). Quartz occurs in igneous rocks such as granite; the quartz crystals in such rocks may weather out to form sand grains and ultimately be redeposited and cemented together to become sandstones.

FELDSPAR: this is the family name for a large group of minerals common in igneous rocks. They are silicates, combining silica with aluminium (A1) and potassium (K), sodium (Na), calcium (Ca), or a mixture of these. Feldspars have a hardness of about 6 and have two good cleavages at, or nearly at, 90° to each other. Orthoclase feldspar ($KAlSi_3O_8$) forms good crystals that are often rectangular in cross-section, although their form is in fact described as monoclinic. They may be cream, grey or pink in colour. Plagioclase is a very variable mineral combining sodium and calcium with silica.

AMPHIBOLES AND PYROXENES: these are dark green, brown or black minerals often with rather elongate crystals that have a six-sided cross-section. They are very complex silicates of iron, magnesium and other elements. (Augite, a pyroxene, has two cleavage planes at about 90°; hornblende, an amphibole found in both igneous and metamorphic rocks, has planes that cross at 56° and 124°.)

MICA: this also occurs in both igneous and metamorphic rocks and is another aluminium silicate. The crystals are hexagonal in outline. Mica has the most perfect cleavage known, cleaving into remarkable thin transparent flakes. It is very soft, the hardness is about 2.5, and can usually be scratched with a finger nail.

CLAY MINERALS: these are the fine-grained constituents of rocks such as clays and shales; they also occur in soils. The individual plate-like mineral particles are so small that they can only be seen with the aid of an electron microscope. They vary widely in colour and usually contain some water. Examples include kaolinite and illite.

CALCITE (calcium carbonate ($CaCO_3$)): this occurs commonly as the skeletal material in organisms, and so it is not surprising that rocks formed largely of the remains of creatures such as corals—limestones—should consist mainly of the mineral form of this compound. It also occurs quite widely in veins intruded into other rocks. The crystals are hexagonal and have three cleavage planes at 75° to each other so that calcite fragments form little 'rhombs'. Calcite is transparent, or nearly so, in its pure form; it also shows the phenomenon of 'double refraction' — when an object is examined through a piece of calcite crystal a 'double image' is seen.

Igneous rocks and structures

The basic classification of rocks into igneous, sedimentary and metamorphic types has already been made, but these categories may be further subdivided.

Igneous rocks, those that solidified from hot, fluid magma (a silicate melt) may be plutonic (Pluto, god of the underworld) if they originated deep within the earth; they are described as hypabyssal if they cooled at an intermediate depth and volcanic if the liquid material reached the surface before cooling. These three categories of igneous rocks can generally be identified by the size of the crystals of which they are composed.

Plutonic rocks have large crystals, 1-5mm across, having cooled very slowly. A typical example is granite, a coarse-grained rock with crystals of quartz, feldspar and mica, having a speckled grey or pinkish appearance. It often formed when large masses of 'acid' (i.e. silica - rich) magma were intruded during the formation of mountain ranges. These large intrusions are known as batholiths and may be many miles in length. Although they form deep within the crust, they may be subsequently exposed by erosion. An example of a granite batholith can be seen in south-west England; the upland masses of Dartmoor, Bodmin Moor, Hensbarrow and Carnmenellis are part of a vast batholith, imperfectly exposed, extending some 220 kilometres (136 miles) from Devonshire south-westwards to the Isles of Scilly. Other large granite intrusions contribute, by reason of their resistance to erosion, to the grandeur of the Highlands of Scotland, e.g. in Inverness-shire and Aberdeenshire. Granite was formerly quarried extensively as a building stone.

Hypabyssal rocks occur in smaller masses such as dykes—discordant, often near-vertical, wall-like intrusions. Dykes are usually, although not always, more resistant than the surrounding 'country rock', e.g. the North Star Dyke, County Antrim, Northern Ireland (see figure 2 and plate 8). They also occur in vast swarms on the Isle of Arran, off the west coast of Scotland. Sills are tabular masses of intruded igneous material concordant with the country rock. Close to the North Star Dyke on the County Antrim coast is Fair Head, a cliff some 194 metres (636 feet) high, cut partly into a sill of dolerite. This is quite a common hypabyssal rock and the Great Whin Sill in Northern England is made of similar material. In Northumberland this sill has been tilted, and so now forms a ridge-like feature. It was along this ridge that the emperor Hadrian built his wall in about A.D. 122. Dolerite is much more finely grained than a plutonic rock such as granite—although the individual crystals are usually visible — and it is less 'acid', so that quartz and feldspar are less in evidence and the darker minerals give it a characteristic heavy grey colour.

Volcanic rocks form when magma reaches the surface to become lava. A common rock of this type is basalt, a very fine-grained, black rock composed of near-microscopic crystals of plagioclase feldspar, pyroxene and a greenish mineral called olivine — but no quartz. Some basalts show a well-developed system of vertical cracks or joints, so that regular hexagonal (six-sided) columns form (plate 7). The best examples in the British Isles are at the Giant's Causeway in Northern Ireland and at

8

Fingal's Cave on the Isle of Staffa. These basalts are the result of a phase of volcanic activity that affected the British Isles about forty million years ago. Further discussion of volcanoes follows in chapter 2.

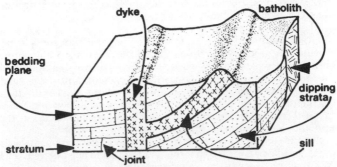

Fig. 2. Some sedimentary and igneous structures.

Sedimentary rocks and their arrangement

Material eroded from the rocks of the continents may be deposited in lakes or seas, the sediments eventually being consolidated by the weight of overlying material. The individual particles may become cemented together by minerals such as calcite. Such rocks show a characteristic layering or 'bedding', the surfaces that separate the individual layers or 'strata' being known as bedding-planes. These water-laid rocks, together with deposits laid down by the wind or by glaciers and rocks such as limestones that accumulate as the result of the activities of organisms, are known as sedimentary rocks.

Conglomerates are rocks composed of pebbles, usually water-rounded and several centimetres in diameter, in a matrix of much finer material; they often superficially resemble concrete and may represent ancient beaches or the deposits laid down by torrents in semi-arid environments (see plate 4).

Sandstones are much more finely grained, as the name implies. Sandgrains about 0.5-2mm in diameter are cemented together by quartz or calcite. Typically sandstones are yellow or white, but a coating of iron oxide round the grains may impart a reddish or brown colour to the rock. Many layers of the Millstone Grit, a sandstone that makes up much of the Pennines, are so coloured. Some sandstones are composed of very rounded grains, which

9

suggests that the particles were blown about by the wind before being deposited in desert dunes. The rock on which Nottingham Castle is built is a desert sandstone of this type. In some sandstones which are either wind-deposited or laid down under the influence of strong currents in a delta or river-mouth 'cross-bedding' occurs. Here some laminae (thin layers) within a stratum are set at an angle to the bedding-planes (see plate 5).

Clay is made up of much finer particles; individual fragments must by definition be less than 0.002mm in diameter. While marine sandstones were probably the result of deposition quite close to the seashore, clay particles are capable of being carried much further out; some clays may thus represent quite deep-water deposits. The typical colour of a clay is blue-grey, but there is considerable variation through the presence of impurities. Clays are easily eroded by streams and commonly floor the valleys of the Midlands and southern England.

Pressure on clay may result in 'lamination', the arrangement of the plate-like particles so that they are parallel to one another. The rock then becomes fissile, splitting very easily, and is known as a shale.

Sedimentary rocks composed largely of calcium carbonate ($CaCO_3$) are known as limestones. Generally such rocks are formed partly from the skeletons of animals such as sea-lilies (crinoids) or corals, and often fossils are clearly visible. Recrystallisation of the calcite, however, may destroy some of the pre-existing structures. Mountain limestone, forming the hills of the Mendips and the Yorkshire Dales, is more resistant than the Chalk of East Anglia and Southern England, the latter being formed partly from the skeletons of microscopic creatures. Oolite (Greek *oon*, meaning egg) is made up of rounded grains of calcite about 2mm across; it is the principal rock-type of the Cotswold Hills.

Metamorphic rocks

Some alteration is involved in the formation of a sedimentary rock from the unconsolidated material; a clay may be subjected to the pressure of the overlying sediment and converted to a shale, a limestone may be partly recrystallised. 'Metamorphism' is the term given to the more substantial alteration of the rocks by heat and pressure. Limestone is completely recrystallised, if heated to a high enough temperature, and marble will form. If a sandstone is similarly heated the individual grains may fuse together and a very hard quartzite will result. Further pressure on a shale, together with an increase in temperature, may lead to the formation of a slate and ultimately a schist (plate 3). Mica is a common constituent of schistose rocks such as those of the Highlands of Scotland, and gives them a distinctive glistening, flaky appearance.

Fig. 3.   Lateral pressure may result in the formation of folds; more intense pressure may result in recumbent folds.

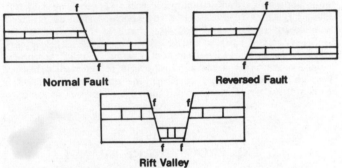

Fig. 4.   Faulting, showing the movement along the fault planes — ff.

Fig. 5.   Wrench faulting, where the movement along the fault plane is horizontal rather than vertical.

Folding and faulting

The occurrence today of rocks such as limestones and sandstones, that originally formed at the bottom of the sea, at heights of 500-800 metres (approximately 1,500-2,500 feet) above sea level in the Pennines and elsewhere provides clear evidence of uplift and earth-movement.

Lateral pressure may result in the formation of upfolds or anticlines and downfolds — synclines — in which strata dip away from or towards a central axis. More intense pressure may result in the formation of recumbent folds. These features are illustrated in figure 3 and plate 1.

A fault is a dislocation or fracture in the rocks along which appreciable displacement has taken place. Tension in the earth's crust may result in the formation of a 'normal' fault (figure 4); compression may result in a 'reversed' fault (plate 2). Where two faults occur in parallel, causing the rocks between them to subside, a rift valley forms. Such is the Central Valley of Scotland, bounded by the Highland Boundary Fault to the north and the Southern Upland Fault to the south. Both can be clearly picked out in the landscape, for along them resistant, intensely crumpled ancient rocks come against more gently folded sedimentaries. 'Wrench' faults are those in which the principal movement is horizontal rather than vertical (figure 5).

2. THE PROCESSES OF CHANGE

'Every mountain and hill shall be laid low, . . . and the rough places plain.'
                                                                Isaiah xl, 4.

Weathering

The 'everlasting hills' are far from eternal, for the surface of the earth is subjected to continual change.

The rocks of the earth's crust are, for example, constantly being affected by the processes of weathering. In upland areas water enters chinks in the rock; this expands on freezing, exerting considerable pressure on the surrounding material, perhaps causing fragments to fall off and accumulate at the bottom of the slope as a scree, as, for example, in the spectacular piles of broken rock on the south-eastern side of Wastwater in the English Lake District. Water also has a chemical effect, attacking the feldspars and micas in a granite and thereby occasioning the disintegration of the rock.

This chemical weathering may have been more vigorous in the past, during a former warmer, wetter period. On Dartmoor and Bodmin Moor, for example, weathering may have proceeded to a depth of several metres, particularly along fissures and joints, causing 'deep-rotting' of the granite. During the cold period that

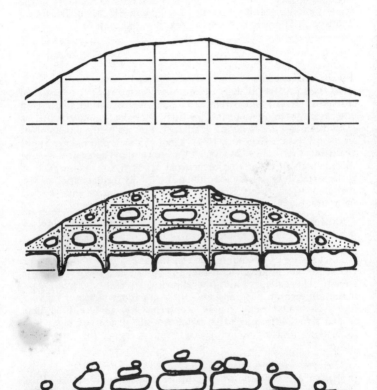

Fig. 6. *The formation of tors in granite country: top, original jointed granite batholith; middle, deep weathering penetrates joints during tropical period and granite disintegrates to leave core-stones; bottom, weathered material is stripped off in cold period to leave up-standing tors.*

13

followed this warm phase, soil-creep or 'solifluction', the accelerated movement of surface material downslope, as the result of the existence of a frozen layer in the subsoil, caused the stripping of the weathered granite to reveal core-stones, the unaltered portions of the granite blocks, piled into irregular castle-like 'tors'. Such features are characteristic elements in the scenery of the granite uplands, e.g. Haytor Rocks on Dartmoor and Rough Tor near Bodmin (see figure 6 and plate 6).

Rivers and streams

At the foot of a hillslope the weathered material is presented to a river or stream and is removed — in suspension, in solution in the case of soluble materials, or by being rolled along the bed of the stream as the 'traction load'. As well as being an agent of transportation, a river may be an agent of erosion in its own right, removing the bed or banks of the stream by corrasion — the mechanical abrasion of the bedrock caused by the impact of transported debris — by solution, or by the hydraulic action of the turbulent stream itself. Rounded potholes may be cut into the bed of a stream.

Most streams tend to meander, or wander from side to side, as they flow downhill; the mechanism of meander development is not understood, but certain observations may be made. The characteristics of meanders, such as distance apart and radius of curvature, are relatively constant in a given stretch of a river or stream. The extent of the lateral swing is relatively small in a mountain torrent, but downstream the meander belt may be very wide. The outsides of bends are frequently undercut so that eroded river cliffs may exist, while on the insides of meanders some deposition of silt may occur, encouraging the growth of rushes.

The cross-sections of river valleys also vary. The V-shaped cross-profile of a mountain stream emphasises the rapid downcutting that is proceeding; along a lowland reach of a river, such as the Severn as it passes through Shrewsbury or the Cuckmere in Sussex, the valley is often much wider and more open.

Some rocks are much more resistant to river erosion than others; where resistant and weak rocks alternate in a gently dipping sequence of strata 'scarp and vale' topography frequently develops. In south Salop steep escarpments of resistant limestone, like Wenlock Edge, overlook valleys floored by more easily eroded shales (figure 7 and plate 10).

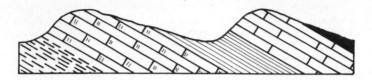

Fig. 7. Scarp and vale topography in alternating weak and resistant strata.

Landscapes of chalk and mountain limestone

The white Chalk of eastern England, the North and South Downs and the Salisbury Plain is very permeable, and so in this downland country surface streams are rare. However, Chalk escarpments are often broken by amphitheatre-like 'coombes' and the more gently sloping back-slopes, such as that inland from West Lulworth in Dorset, may show a complex network of dry valleys. Sometimes these valleys contain flowing streams following a period of heavy rain, but many of them are known not to have contained running water within living memory. Many theories of origin have been proposed: one suggestion is that streams cut the valleys in rainy or 'pluvial' periods in the past; another that they were cut during a cold phase during or following the Ice Age (see chapter 3) when the rock below the surface was frozen and acted as an impermeable stratum, so compelling streams to flow over the surface.

Mountain limestone is much more massive than Chalk and some of the areas which it underlies — the Mendip Hills, parts of the Derbyshire Peak District and Craven, in Yorkshire, as well as a portion of Sutherland in the north of Scotland — form uplands over 350 metres (*c.* 1000 feet) high, while the chalklands of the south are usually at only half that elevation. Dry valleys occur and may be of gorge-like dimensions, e.g. Cave Dale and the Winnats Pass near Castleton in Derbyshire and Cheddar Gorge in Somerset. Such limestones are usually well-jointed. Solution along joints by rainwater in which some small amount of carbon dioxide from the atmosphere has been dissolved frequently results in the formation of a criss-cross pattern of deep clefts in the broad, flat 'limestone pavements' such as that above Malham Cove in Yorkshire. Craggy outcrops of bare rock known as 'scars' are common, for example Giggleswick Scars, not far from Malham. Solution may continue underground and an elaborate network of caverns and tunnels may be dissolved over thousands of years. What happens is that the slightly acidulated rainwater converts the calcium carbonate of the limestone to calcium bicarbonate, which, being soluble, is readily removed. Sometimes the lime in the percolating water is redeposited when water evaporates in a cave, so that the intricate and beautiful forms of stalactites grow from its roof, stalagmites marking the points where water drips on to the cavern floor. Magnificent examples of these, some almost human in their outline, can be seen in many of the caves in limestone areas open to the public, like the Treak Cliff Cavern in Derbyshire and Wookey Hole in Somerset.

Coastal erosion

The hydraulic action of breaking waves at the base of a cliff can produce pressures of the order of 60 metric tons per square metre (over 12,000 lb per square foot) as the air is compressed within fissures in the rocks. Beach material — rock fragments and shingle — may be thrown at the cliff surfaces and weaken them. Wave action also removes the weathering products that fall from the cliff face and brine has a much greater chemical effect on many minerals than fresh water; it has been estimated that it is about fourteen times as effective in dissolving orthoclase feldspar.

A wave-cut notch therefore develops at the base of the cliff and severe undercutting may occur, resulting eventually in the collapse of part of the cliff. This can be seen north of Flamborough Head in Yorkshire and at Hunstanton (figure 14) and Cromer on the Norfolk coast. Many cliffs, therefore, are being actively eroded by marine attack. The rate depends on the rock type, being particularly rapid for cliffs of glacial till (see chapter 3) in eastern England. A rate of 2 metres (6.5 feet) per year has been noticed on the Holderness coast. Old maps have enabled estimates to be made of the rate of erosion in the past. At Dunwich, on the Suffolk coast, for instance, the shore seems to have advanced landwards about 400 metres (1300 feet) since 1578 and over 150 metres (500 feet) since 1754.

Wave attack may work on lines of weakness, faults and major joints for example, to form marine caves and natural arches like the Green Bridge of Wales in south Dyfed and Durdle Door near Swanage. Small portions of exposed headlands may be cut off — the roof of a natural arch may collapse — to form 'stacks' such as Old Harry, the isolated pinnacle of Chalk that stands in the sea off the Isle of Purbeck, or The Needles across the Solent at the western tip of the Isle of Wight. The wave-cut platform that slopes away from the base of the cliffs on such exposed coastlines is usually covered by large boulders, the finer materials having been swept away. Sheltered bays generally have a more gently sloping profile, cliffs are lower and the beaches are covered with finer, sandy material.

Deposition

Material eroded by rivers and by wave attack is, of course, eventually deposited elsewhere. Silt and clay particles may accumulate where the velocity of a river is reduced as it enters the sea to form a delta, so called because of its outline's resemblance to the Greek letter of that name. Mudflats form and are colonised by salt-marsh plants; these help to stabilise the unconsolidated material and accelerate upward growth.

Beach material, such as sand and shingle, may be moved along the shore by 'beach drift'. This process operates when waves generated by the dominant wind strike the shore at an angle; the 'swash' — the water from the breaking wave — throws shingle diagonally up the beach, but the 'backwash' will drag pebbles back down the steepest slope. Where a coast changes direction, at a river mouth or bay, for instance, a spit of sand and shingle is often built out, continuing the line of movement. For example, westward-pointing spits have developed at Hurst Castle and Calshot on the Solent under the influence of waves generated by the prevailing winds from the south-west. Sometimes rivers may be diverted some distance by such spits — the river Alde is deflected 18 kilometres (over 11 miles) southwards by Orford Ness. Here the south-west wind, blowing seawards off the coast of East Anglia, has little effect; the dominant spit-building waves are those from the north-east.

Some material is carried a fair distance from the coast. Rivers such as the Rhine, Thames and those of East Anglia bring many thousands of tons of material into the southern North Sea each year and one estimate has it that the sea floor is subsiding at the rate of about 2mm (0.08 inches) per year, partly as the result of the weight of accumulating sediment. The sand and mud deposited in such environments, in partly enclosed seas and on continental shelves, following uplift, will constitute the sandstones and clays of the future.

The movement of crustal plates

Changes are also occurring on quite a different scale. It is now believed that the earth's crust is made up of six major 'plates' of crustal material and a number of smaller ones. In places these are moving apart; thus the Eurasian plate, made up of Europe, Asia and the floor of the eastern North Atlantic, is moving away from the American plate. Along the Mid-Atlantic Ridge, therefore, earthquakes are not infrequent, and there are a number of volcanic islands, the largest of which is Iceland, which formed as the result of the upwelling of material from below along the rift created as the plates moved apart at an average rate of a centimetre or two a year. This gradual movement is sufficient to explain the separation of Europe from North America and of Africa from South America over the last seventy million years. There are rocks in Scotland that resemble some of those in Labrador and Greenland; there are structures in south-west England that reappear in Newfoundland. It now seems likely that North America was once connected to Europe and that the North Atlantic has formed and widened over the recent geological past. Volcanic material emerged from the rift accreting on to the plate margins as they parted, forming the floor of the expanding Atlantic Ocean.

Other types of plate-margin can be seen elsewhere in the world. Off the coast of eastern Asia the result of the movement of two plates towards one another can be studied; one plate is buckled up to form chains of islands and mountains such as those of Japan, the other is dragged down beneath the upper plate causing deep ocean trenches to form. It is now widely held that the convergence of the crustal plates and the squeezing and crumpling of the material deposited near their margins is the principal cause of mountain building. The margin between the American and Pacific plates follows the line of the San Andreas Fault in California. Along this wrench or tear fault the two plates are shearing against one another as they move in more or less parallel but opposing directions.

Volcanoes and earthquakes

There are about 600 volcanoes which are either active or have not been dormant for sufficiently long to be deemed extinct. Most of them occur at or near the boundaries between the main crustal plates, where the crust is unstable. Off the south coast of Iceland, in November 1963, a new island, Surtsey, appeared. A narrow ridge became a cone 150 metres (about 500 feet) high, composed of volcanic ash and cinders, within a month. Six months after it first appeared fluid lava began to flow and the island remained active for several years. A few kilometres from Surtsey another volcano on the main island of the Westman group began to erupt in January 1973 after a period of dormancy lasting several thousand years.

Some eruptions have been explosive in their violence and very destructive. In 1883 two thirds of the island of Krakatoa in Indonesia was destroyed by a volcanic explosion and 36,000 people were killed, most of them by the tidal wave that followed the eruption. In 1902 30,000 were killed by the eruption of Mont Pelée in Martinique in the West Indies.

A volcano is usually conical in form, but the angle of the sides of the cone depends on the composition of the lava. Acid lava, which contains a lot of silica (SiO_2), tends to be treacly and slow-moving; the cones it forms are thus steep-sided and almost bulbous in shape, while more basic lavas (lacking silica) are more mobile and form cones with gentler gradients. A volcano is subjected to weathering and erosion and the outer layers of lava may eventually be removed to reveal a volcanic 'neck' or 'plug', the crater-infill, often composed of more resistant rock than the surrounding lava, as an upstanding pinnacle. Such are the rocky crags on which Edinburgh and Stirling castles are built.

Besides the volcanic rocks of County Antrim, described in chapter 1, which were extruded through a number of vents now plugged by igneous masses rather like smaller versions of these Scottish examples, and which form steep-sided hills in the Ulster landscape, extrusive rocks appear in the Hebrides and in Cumbria. There were volcanoes on the islands of Mull, Skye, Rhum and on Ardnamurchan 50-75 million years ago, and the jagged hills of many of the islands of Scotland are carved from volcanic rocks. The Borrowdale Volcanic Series of the central Lake District are much older and provide evidence for crustal instability about 500 million years ago.

Earthquakes are caused by rock masses moving against one another within the earth, relieving the strains and stresses that build up from time to time in the planet. Only a small proportion of the shocks are severe enough to be noticeable; many can only be detected by sensitive instruments called seismographs. The 'epicentre' of an earthquake is the point on the earth's surface vertically above the 'focus' or origin. If the epicentres of a large number of earthquakes are plotted on a map, many will be found to lie along the margins between the crustal plates, the zones of crustal instability where one would expect movement to occur. The earthquake zones and major volcanic belts are thus approximately coincident with one another.

Yellowstone Park and adjacent parts of Montana and Wyoming in the U.S.A. are in an area where volcanic rocks are widespread but where present-day activity is limited to geysers and hot springs. This area was affected by a very severe earthquake on 17th August 1959; rockslides occurred (one damming a valley to form a lake), buildings were damaged, metal fences were buckled and fissures — some with a vertical displacement of several metres — appeared in roads. The area is one of low population density and casualties were few, but where earthquakes occur in built-up areas the results are catastrophic. In the Sagami Bay earthquake of 1927 much of Tokyo was destroyed and 140,000 people were killed; 14,000 people perished at Agadir in Morocco in 1960.

Because of its position in the heart of one of the stable crustal plates earthquakes of any magnitude are now rare in Britain, but small disturbances do occur. In August 1965 a minor earthquake with its epicentre off the coast of Suffolk caused the ground to shake and windows in buildings to rattle along the coastal region of East Anglia. Other tremors have been noted along the major fault lines in Northern England and Scotland.

3. THE GEOLOGICAL HISTORY OF THE BRITISH ISLES

The determination of the ages of rocks

There are a number of ways in which rocks may be dated. The simplest is the 'Law of Superposition' which states: 'Provided they remain undisturbed, in a succession of horizontally bedded rocks, the oldest will be found at the base and will be overlain by the younger ones'. Fossils (the remains or traces of past life preserved by natural processes in the rocks of the earth's crust — see chapter 4) are also important; where two rocks contain the same assemblage of fossils one may presume that they were deposited at the same time. There are, however, difficulties in the use of fossils for dating. Two rocks may be contemporaneous (laid down at the same time) but contain the remains of different animals and plants because they were deposited in different environments. One would not expect deserts, coral reefs, mangrove swamps and shallow estuaries to be inhabited by the same kinds of creatures today, so it is not surprising that sandstones, limestones, coal deposits and siltstones, for instance, contain different fossils. On the other hand, some organisms existed for immense periods of time and so are of little use for dating; the little shelled creature *Lingula* was living when rocks nearly 600 million years old were being deposited and it still survives. Ideally, 'zone fossils' are used; these are creatures that were once common and widely distributed, preferably living in a range of different environments, yet only persisted for a comparatively short period. A zone fossil

20

characterises a particular layer of rock or 'stratigraphical zone' (stratigraphy is historical geology, the branch of the subject that attempts to unravel complex changes that have occurred throughout geological time).

More sophisticated methods exist. When igneous rocks solidify, the crystals of which they are composed contain small quantities of radioactive elements such as uranium and thorium. These are gradually and at a constant rate changing to stable, non-radioactive elements — in the case of the two examples mentioned, to lead. If a sample of an igneous rock is analysed very accurately the proportion of uranium to lead can be determined and as the rate of 'radioactive decay' is known the time that has elapsed from the original crystallisation of the rock from magma can be estimated.

With more recent deposits a range of techniques is available. Calcium phosphate in bone absorbs fluorine from percolating water at a constant rate forming a mineral called fluorapatite. A determination of the quantity of this mineral in bone material can enable an estimate of the age of the bone, and thus, perhaps, of the surrounding sediment, to be made. In 1911 a remarkable skull, similar in some respects to modern man but with an ape-like jaw, was discovered at Piltdown in Sussex. Partly as the result of the determination of the fluorine content it was shown that while the upper part of the skull was about 50,000 years old the jaw was from a modern ape. The two had been brought together from elsewhere and placed in a gravel bed, presumably as a hoax.

Cosmic rays continually bombard the upper layers of the atmosphere from outer space and in so doing convert a small amount of the nitrogen in the air to radioactive carbon-14, some of which, after it has become mixed into the lower layers of the atmosphere, is absorbed by growing plants. Carbon-14, like other radioactive matter, eventually decays to a stable, non-radioactive form; when a plant dies assimilation of carbon-14 ceases and so the amount present in a sample of wood or peat can reveal how long it is since the plant that produced it was alive.

By a careful study of the layers of rock in natural exposures, quarries and boreholes together with the comparison of one sequence of rocks with those from other localities, the analysis of the fossil content of the rocks and the use of some of the more sophisticated scientific methods described above, a 'geological column' has been built up. Strata have been grouped into 'systems', the rocks in one system being those laid down in one 'period'; periods are then grouped into 'eras':

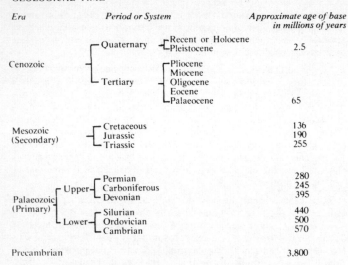

Era	Period or System		Approximate age of base in millions of years
Cenozoic	Quaternary	⌈Recent or Holocene ⌊Pleistocene	2.5
	Tertiary	⌈Pliocene │Miocene │Oligocene │Eocene ⌊Palaeocene	65
Mesozoic (Secondary)	⌈Cretaceous │Jurassic ⌊Triassic		136 190 255
Palaeozoic (Primary)	Upper⌈Permian │Carboniferous ⌊Devonian		280 245 395
	Lower⌈Silurian │Ordovician ⌊Cambrian		440 500 570
Precambrian			3,800

The reconstruction of the past

'The present is the key to the past' wrote one early authority, for the determination of past conditions often depends upon a comparison of rocks deposited long ago with modern environments. Corals today inhabit warm, shallow seas and we may assume that certain limestones, which are in effect fossil coral reefs, were built up under broadly similar conditions. In ice caves beneath glaciers in Iceland and Greenland a characteristic unsorted material, a confused mixture of particles ranging from large boulders to minute specks of rock-flour, can be seen actually being carried along by the slowly moving ice. Similar material, known as glacial till, covers much of East Anglia. Today we infer that during the Pleistocene, the Ice Age, glaciers and ice-sheets covered much of Britain. From a detailed study of rocks, the materials of which they are composed and the fossils they contain it has been possible to determine the position of coastlines 450 million years ago. And by studying the way in which lime was deposited to form the skeletons of fossil creatures it has proved possible to suggest what the mean temperature of the sea was within 1°C in remote geological periods. Geologists have even been able to estimate the salinity of former seas and to say something of the tidal conditions that existed as long ago as the Devonian period.

Sometimes studies of this sort reveal something quite unexpected. Modern corals develop daily growth bands in their skeletons as a result of the regular rise and fall of the tide. The nature and rate of deposition of this limy skeletal material varies between summer and winter and so a seasonal pattern is superimposed on the daily one; thus the whole life of a coral may be chronicled. Very similar layering occurs in some fossil organisms and a count of the growth bands in some Devonian corals has indicated that a year 375 million years ago consisted of about 400 days and thus that the rotation of the earth is gradually slowing down.

The oldest rocks in Britain

Rocks representing the most recent of the geological systems are found in south-eastern England and one tends to pass over older rocks as one moves towards the north and west of the country. Precambrian rocks, deposited more than about 600 million years ago when animals with hard parts likely to survive as fossils were rare or entirely absent, are thus to be found mainly in north-west Scotland, Ireland, Anglesey and North Wales. There are also a few 'inliers' of Precambrian rocks in the Midlands, Yorkshire and the Welsh Borderland (an inlier is an outcrop of old rock surrounded by younger strata).

The most ancient rocks in Britain are those of the Isle of Lewis and nearby parts of the mainland of Scotland, for example near Scourie and Drumbeg in Sutherland. Before the most recent separation of the continents these Lewisian rocks were probably part of the Canada-Greenland mass. They are devoid of fossils, intensely folded, metamorphosed and intruded by countless dykes. The oldest rocks of all date from more than 2,600 million years ago and are banded pyroxene and hornblende granulites (granulites represent the most extreme form of metamorphism and develop under conditions of very high pressure and temperature). Some are intricately folded, suggesting that the sequence represents at least two, possibly three, mountain-building periods.

Other important ancient sequences of rocks in the Central and Northern Highlands of Scotland, in Northern Ireland (e.g. on the County Antrim coast near Ballycastle) and Donegal, less old than the Lewisian but probably Precambrian in age, are the Moinian and Dalradian. The original sedimentary nature of these rocks is generally discernible, although they too have been metamorphosed. They frequently consist of mica-schists and quartzites.

Here and there in the north-west of Scotland, from Cape Wrath in the north southwards to Loch Torridon and beyond, at times forming steep cliffs overlooking the Minch, in places forming bare mountains, generally pyramidal in shape, are Precambrian rocks of a rather different sort. The Torridonian Sandstone Series was probably deposited about 850 million years ago; it is thousands of metres in thickness, bright red in colour and contains feldspar fragments. This suggests desert conditions, although the rainfall may have been rather higher than that received by some modern deserts. In the total absence of plants (land plants did not appear until many millions of years later) wind erosion must have been much more significant and some areas that had appreciable rain would have appeared as desert-like landscapes.

Rather younger are some of the Precambrian rocks of Anglesey. Here, in the Mona series, are quartzites and slates as well as some volcanic rocks. Besides being metamorphosed they have been much affected by folding and on Holy Island, for example, complex contortions can be examined. Rocks called 'greywackes', a coarse type of green-grey slate, thought to be of similar age to the Mona rocks of Anglesey, are found at Ingleton in Yorkshire; they are worked as a source of road-building material.

In the hills of Salop ancient volcanic rocks have been found and the Longmynd is made up of Precambrian sandy sediments. The Charnwood Forest and the Malvern Hills are similar inliers of resistant ancient rocks surrounded by lowland of much more youthful sediments.

The Lower Palaeozoic and Caledonian Mountain Building

Whereas the Precambrian rocks described above are un-fossiliferous, fossils begin to appear in large numbers at the base of the Cambrian (the name of the system comes from the name for Wales in the Welsh language). Some animals and plants must have lived in Precambrian seas (indeed very simple fossils have been found in African rocks nearly 3,000 million years old), but quite suddenly, about 600 million years ago, there appeared a number of groups with hard shells or skeletons. These included the trilobites and brachiopods like *Lingula*.

Two extensive areas of Cambrian rocks appear in Britain, one in North Wales, the other in the north-west of Scotland. They differ markedly from one another, both in rock-type and in the fossils they contain.

The rather desolate tract of country known as the Harlech Dome in Gwynedd is made up of a great thickness of Cambrian rocks — many thousands of metres of sandy deposits such as flagstones and grits which probably accumulated in a narrow, rather deep, trough-like sea or 'geosyncline' as the result of the vigorous erosion of the surrounding land areas. The presence of

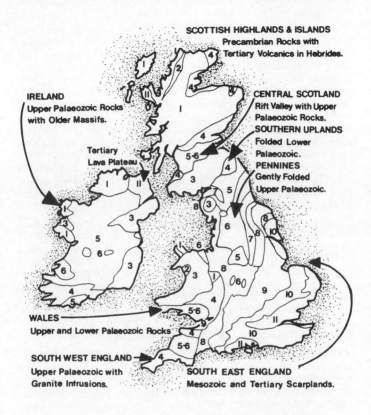

Fig. 8.   A geological sketch map of the British Isles:

- 11 Tertiary
- 10 Cretaceous
- 9 Jurassic
- 8 Triassic
- 7 Permian
- 6 Upper Carboniferous
- 5 Lower Carboniferous
- 4 Devonian
- 3 Ordovician, Silurian
- 2 Cambrian
- 1 Precambrian

appreciable percentages of manganese in some layers of shales has been held to suggest that some of the nearby continental areas were subjected to vigorous tropical weathering.

In north-west Scotland, for example near Durness in Sutherland, Cambrian rocks that formed in quite a different environment appear. Here are conglomerates and limestones with some sandy layers that were deposited in a shallow sea — the presence of worm-burrows and ripple-marks suggests this. The fossils are quite unlike those of the Welsh Cambrian only 300 miles away, and are in fact much more similar to those found in strata of approximately the same age in east Greenland and North America — which were then probably quite close to Scotland. The limestone areas near the head of Loch Assynt and round the Kyle of Durness and Loch Eriboll form stretches of lower, rather greener farmland in sharp contrast to the bleak uplands of crystalline rocks nearby. In places, however, the limestone reaches the sea in spectacular cliffs with caves, like Smoo Cave, and sea-stacks.

The rocks of the Ordovician — so called after an ancient British tribe—in many respects have affinities with the Cambrian strata of Wales. In North Wales — Snowdonia and the Lleyn peninsula — thick layers of shales were laid down in the trough-like basin of deposition that persisted from the Cambrian. Amongst the commoner fossils are graptolites such as *Didymograptus*. There are differences however; while the Cambrian was a relatively tranquil period, volcanic activity and earth movement were much more evident in the Ordovician. Snowdon was a centre of activity and lavas and ashes 600 metres (2,000 feet) in thickness appear.

Ordovician rocks also make up a substantial proportion of the Lake District; here the volcanics are even thicker and the Borrowdale Volcanic Series gives the central core of the massif, the area round Helvellyn, much of its wild, craggy character. Further north, although Skiddaw is 930 metres (3,053 feet) high, the mountains are more even in outline and bare rocky crags are rarer. Here the rocks are highly folded and cleaved slates of early Ordovician date.

Tightly folded Ordovician rocks are also seen in the Southern Uplands of Scotland and in County Down and elsewhere in Ireland.

The Silurian system was named after the Silures, a Welsh border tribe. Silurian rocks underlie more than half of Wales, occurring in a broad crescent centred on the Cambrian Harlech

Dome in North Wales and forming the main mass of the mountains of Central Wales.

Sedimentation in the trough or geosyncline continued; deposits are mainly deep-water sediments — slates and shales containing graptolites — in Central and West Wales, shallow-water limestones appearing further east in Salop. These limestones dip sharply to the east and as they alternate with shales give rise to a 'scarp and vale' topography, seen best at Wenlock edge (plate 10). The limestones here have a prolific fauna and a wide variety of trilobites, brachiopods, corals and crinoids (sea-lilies) are to be found. Reef-like structures interrupt the normal bedding in places — localities where there was vigorous upgrowth of coral from the sea-bed. These same strata reappear in south Staffordshire, where the trilobite *Dalmanites* is so common that it has locally been given the name 'Dudley locust'!

During the Cambrian, Ordovician and Silurian a layer of sediments up to 11 kilometres in thickness had accumulated in the trough (geosyncline) that extended across the British Isles. These rocks were then crumpled by compressional forces and a range of mountains, the Caledonian Mountains, was thrown up. They ran southwards along the length of Scandinavia, through Scotland, Wales and Ireland and into a then much closer Newfoundland and Appalachia. Folds and thrusts formed, granite batholiths were intruded — in the Lake District for example — and metamorphism took place on a widespread scale, as in the formation of the slates of North Wales.

The Caledonian Mountains were not unbroken; depressions, basins of inland sedimentation, existed in north-east Scotland — the Orcadian Basin — and in the then recently formed rift valley that existed across Central Scotland.

In these depressions iron-stained red sandstones of immense thickness were deposited; some 350 metres of sandstone (over 1,000 feet) are exposed in sea cliffs on the Isle of Foula and in the Orkneys. The occurrence of breccias and conglomerates suggests the rapid erosion of nearby uplands. The local abundance of unweathered feldspathic sand-grains suggests an arid climate, but lacustrine rocks containing fish fossils also exist, e.g. in Caithness. Suncracks in some of the mudstones, however, imply that at least some of these lakes intermittently dried up.

Volcanic activity continued in Scotland and northern England. The Cheviot was an active volcano; the Cheviot Hills today consist of lavas arranged concentrically round a granite intrusion and traversed by a series of dykes.

The Upper Palaeozoic, Variscan Mountains and New Red Sandstone deserts

The continental rocks of this period, described above, are collectively known as the Old Red Sandstone (ORS) to distinguish them from rocks deposited in marine environments in the south of the country, the Devonian rocks. Green, brown and purple shales and sandstones were deposited in north Devon, but limestones, often pink in colour and attractively marked on account of the coral fossils they contain, appear at Plymouth and Torquay.

These marine deposits were the first to be laid down in a new geosyncline, the main trough of which ran east-west; for much of the Carboniferous — the system that followed the Devonian — two discrete areas of sedimentation existed, one in northern England and southern Scotland, the other in South Wales and the southern part of England, separated by a ridge of land — 'St George's Land' — extending east-west from south-eastern Ireland across Wales and the Midlands towards the continent.

The sea that lapped round this land area was for the most part warm, clear and shallow; corals, sea-lilies and brachiopods flourished and thick limestones were deposited. Great thicknesses of limestone can be seen in a continuous section in the Avon Gorge, near Bristol. Further north, the Great Scar Limestone in Yorkshire and the Melmerby Scar Limestone of the northern Pennines give rise to the 'mountain limestone' topography already described. But in general as one moves northwards the massive limestones are seen to give way to an alternating succession of thin limestones (up to about 10 metres in thickness), shales, sandstones and coal seams. This succession of rocks, the Yoredale series, shows 'cyclic' sedimentation, each unit of strata averaging perhaps 25 metres (about 80 feet) in thickness and being known as 'cyclothem' (figure 9). The sequence of different beds within a cyclothem shows shallow clear seas gradually being silted up as material from nearby land was brought in until the mudflats and sandbanks emerged above sea level and soils and forest cover developed. Eventually submergence occurred and the cycle started again; it was repeated many times. The miners of former centuries gave individual limestones descriptive names such as the Five Yard, Four Fathom and Cockleshell. The alternation of rocks of differing resistances gives many hillsides in the north of England a distinctive stepped profile. In places in Teesdale, for example, the limestones stand out as miniature escarpments, layers of shales being eroded to form more gently sloping terraces.

Above the massive limestones in Derbyshire and Yorkshire occurs the Millstone Grit — so called because of its former use for making stones for grinding flour. This is a coarse-grained sandstone, in places 2,000 metres (6,500 feet) thick, sometimes red in colour when fresh but weathering to a very dark hue. It was deposited in a delta, a fan-shaped mass of material brought down into the sea north of St George's Land by a river from a land-mass occupying the present position of Scotland. In places there are conglomeratic layers but shale bands also occur. Fossils are rare. It forms the moorlands of Bleaklow and Kinderscout, standing out as sharp, scarp-like 'edges' on the sides of the valley where rivers have cut into the tabular uplands.

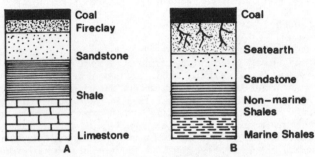

Fig. 9. Cyclothems from: A — Yoredale Series; B — Coal Measures. There are considerable variations but these sequences are typical.

The massive sandstones of the Millstone Grit give way vertically to the Coal Measures, a cyclic succession like the Yoredales (figure 9B). The coal seams, which vary in thickness from 2 to 3mm to about 2 metres, make up some 4 per cent of the total thickness of the succession, separated from one another by shales, siltstones and seatearths — fossil soils. Although coal seams are by no means confined to it, the name of the Carboniferous system means 'coal-bearing'. Coal formed from the impeded decomposition of vegetable matter in waterlogged conditions. Probably submergence and the covering of the dead vegetation by sediment followed the growth of sub-tropical forests of fern-like plants in low-lying swampland. These plants (e.g. *Calamites* and *Lepidodendron*) are quite common as fossils in the seatearths that are associated with the coals and specimens can sometimes be found on tip-heaps near coal mines.

Another period of mountain building, the Variscan or Hercynian, followed the deposition of the Devonian and Carboniferous sediments. In the southern part of the British Isles — in South Wales, south-west Ireland and in Devon and Cornwall — the foldings had an east-west trend; in Northern England the Pennines were thrown up. The Coal Measures, and often the Millstone Grit below them as well, have been eroded since the uplift, but where the Measures were downwarped or faulted downwards, as in the Lancashire, East Yorkshire, Durham, Northumberland and Nottinghamshire coalfields, the seams have been preserved from denudation.

A number of comparisons may be made between the ORS that followed the Caledonian earth-movements and the Permo-Trias or New Red Sandstone (NRS) that came after the Variscan movements. (The Permian, named after one of the former provinces of Russia, and the Trias, so called because of its threefold division in Germany where early investigations were carried out, are often grouped together, in spite of differences in the fossils they contain.) In both, continental deposits were laid down in arid or semi-arid environments. Sandstones have features suggesting that they consist of former sand-dunes, showing, for example, diagonal 'dune-bedding' and the rounded 'millet-seed' grains that reveal a history of abrasion by wind action. Such are the dark red Penrith Sandstone of north-west England, the Yellow Sands that appear above the Coal Measures of County Durham and some of the Triassic rocks of Nottinghamshire. Conglomerates also occur, the result, perhaps, of intermittent desert torrents from the Variscan Highlands.

Marine conditions extended from northern England towards Germany and Central Europe. The sea, the Zechstein Sea, was subjected to severe evaporation, for besides limestone with corals— the Magnesian Limestone that forms cliffs along much of the County Durham coast — beds of rock salt (Na Cl), anhydrite (Ca SO_4) and potash and iodine salts were laid down. These are worked as sources of raw materials for the chemical industries of Teesside.

The Jurassic, Cretaceous and Tertiary

After this rather long continental phase the mountains had been eroded down to a low level and marine conditions returned with the Jurassic.

The Lower Jurassic, the Lias, is well exposed in the Dorset coast near Lyme Regis and in sea cliffs near Whitby in Yorkshire. Shales with thin limestones occur; at Whitby, Jet, formed from woody material, and fossil remains of stranded reptiles suggest that the waters were shallow (Jet is a black semi-precious stone that takes a brilliant polish and is used locally for making ornaments). The Middle Jurassic, in western and central England, is made up of thick oolites — limestones composed of egg-like grains. It is these that form the steep escarpments of the Cotswold Hills. Further north a delta existed, rather similar to the Millstone Grit delta of the Carboniferous. The sandstones so deposited now constitute the tabular, resistant hill country of the North York Moors. A few poor coals are found and plant fossils — cycads for example—are to be found on the flanks of Roseberry Topping on the northern fringes of the moors. The Upper Jurassic is typified by substantial thicknesses of clays, the Kimmeridge and Oxford Clays, for example, that underlie much of the middle of England. Fossils such as ammonites and bones and teeth from marine reptiles are common in places; quite a good collection may be made from the clay-pits near Peterborough where clay is worked for brick-making.

The infilling of the Jurassic sea with deltaic and estuarine deposits resulted in a gradual shrinkage of the marine areas. At the end of the Jurassic sediments were being deposited over a much reduced area of southern England. The Portland stone, a limestone found on the Isle of Portland well known as a building material, dates from this late stage, as do the Purbeck strata, so well displayed in the cliffs south of Swanage in Dorset. The presence of freshwater deposits and fossils shows that eventually this much reduced area of deposition was cut off from marine influences.

This area of swamps and lagoons, covering southern England from Kent to Dorset and extending as far north as Oxford, continued into the Cretaceous. Here the sands of the Ashdown Forest were laid down, along with clays and siltstones; in some of these dinosaur footprints have been found. Another separate area of deposition existed for part of the Cretaceous in parts of Lincolnshire and Norfolk. Conditions in the two areas differed, for while the thick, stiff, dark Gault Clay was being deposited in the south, its place in the northern area is taken by the Hunstanton Red Rock, or Red Chalk. In the cliffs at Hunstanton it appears as a conspicuous bright red layer, only a metre thick. The red colour is probably due to its deposition adjacent to a land-mass experiencing tropical weathering.

As the Cretaceous wore on the sea once again transgressed across the British Isles until they were almost entirely covered by a shallow, warm sea in which a lime-rich mud was deposited. Probably just the mountains of the Scottish Highlands protruded.

The deposit laid down in this sea was eventually compressed and consolidated to become the Chalk of the dazzling white cliffs of Kent and Sussex and the rolling hills of the Chilterns, the North and South Downs and the Yorkshire Wolds. Flints, nodular masses of almost pure silica, are common in the very uniform Upper Chalk; their mode of origin is still uncertain but they have been worked at or near Grime's Graves in Norfolk for about 4,000 years — by the neolithic folk for axe-heads and more recently as ornamental facing stone.

The covering of Chalk, in places several hundred metres thick, has been removed from almost all its former extent in northern and western England and Scotland. In Northern Ireland it has been preserved beneath a protective capping of volcanic rocks. The juxtaposition of the brilliant white Chalk and the almost black basalt layer above it has led to the rugged cliffs of County Antrim being described as having a 'wild, magpie appearance'.

This, the last marine phase to affect a major proportion of Britain, ended with wide scale uplift and a period followed in which no deposits were laid down. Indeed extensive erosion took place before the south and east of England were once again submerged during the Tertiary.

Eocene and Oligocene (i.e. older Tertiary) sands and clays were deposited over most of East Anglia and southern England from Kent to Dorset, but it is only in the slightly downwarped London and Hampshire Basins that the sediments of the older Tertiary survive. Conditions varied both laterally within the area of deposition and also with time. Both marine and freshwater deposits are represented. The most important layer is the stiff London Clay which is blue in colour when freshly exposed but weathers to a brownish hue. Fossil seeds, fruit and leaves from some five hundred species of plants have been found in this Eocene stratum and their nature suggests that tropical rain forests covered the surrounding land.

This temporary marine depositional phase ended with the Miocene, when southern Britain was again uplifted and further erosion took place.

The Ice Age

Rocks confirmed as being of Pliocene age are rare in Britain, but a small outcrop of Coralline Crag, a yellowish sandy material, is to be found in a narrow strip close to the Suffolk coast between Boyton and Aldeburgh and is almost certainly of Pliocene date. The title 'Coralline' is misleading, for although corals do occur, they are not numerous. It was the presence of a large number of

1. *Folded Carboniferous strata, Loughshinny Bay, Co. Dublin, Eire. Notice the glacial till at the top of the section and also the effects of wave attack on inner parts of the folds at the base. (Height of section about 15 metres.)*

2. *A reversed fault in Rousay Flags (Old Red Sandstone), Icy Geo, Rousay, Orkney. Reversed faults are the result of compression within the earth's crust.*

3. *Folding in striped schists, Loch Eilt-Glenfinnan road, Inverness.*

4. Conglomerates are rocks composed of pebbles in a matrix of finer material. This Portraine Conglomerate (Old Red Sandstone or basal Carboniferous) is at Burrow Townland, near Swords, Co. Dublin, Eire.

5. Cross-bedding in red Melby Sandstone, Sound of Papa, Shetland; some laminae (thin layers) are set at an angle to the bedding planes.

6. *A tor formed as the result of the weathering of granite: Rough Tor, Bodmin Moor, Cornwall.*

7. *Hexagonal columns of basalt, formed through the processes of weathering on a well-developed system of vertical joints, Giant's Causeway, Co. Antrim, Northern Ireland.*

8. *North Star Dyke, near Ballycastle, Co. Antrim, Northern Ireland. The igneous intrusion is more resistant than the surrounding Carboniferous strata and so stands out conspicuously on the foreshore.*

9. *A natural arch in columnar basalt, White Park Bay, Co. Antrim. Note 'raised beach' at cliff foot, formed by fall in sea level.*

10. *Scarp and vale topography, Wenlock Edge, Salop. Resistant Silurian limestones stand out as escarpments while less resistant shales have been eroded by river action.*

11. *The effects of highland glaciation in Cumbria—Blea Tarn and Haweswater.*

12. *Mineral exploitation—china clay workings at St Austell, Cornwall.*

13. *Derelict copper mines, St Day, Cornwall.*

14. *Ammonites, a form of mollusc, in Liassic shales on the foreshore at Portrush, Co. Antrim, Northern Ireland.*

15. *Fossil trees, 'Lepidodendron', Victoria Park, Glasgow.*

Bryozoan fossils, a quite different type of simple animal, that gave rise to the confusion. These and other fossils show that the Coralline Crag sea was relatively warm; indeed some of the organisms found as fossils in this stratum are still to be found alive in the Mediterranean.

The Red Crag and the Norwich Crag which rest on the Pliocene in Suffolk are in some respects similar, although the fossils found in these unconsolidated yellow-red sands are rather different. There is a much higher proportion of cold water forms; some of the species of these Crag deposits are found today living in the sea off northern Norway. This contrast provides evidence of the gradual cooling of the climate in the early Pleistocene.

At times during the Pleistocene or Ice Age the mean annual temperature in Britain must have been over 15°C (27°F) colder than it is at present; the snow-line was at less than 700 metres (c. 2,300 feet) in the uplands and ice accumulated. Ice-scooped hollows (corries, cirques or cwms) are common in such regions; typical examples are the Blea Tarn corrie in the Lake District (plate 11) and Cwm Idwal in Snowdonia. Valleys that guided the paths of glaciers became over-deepened and now show U-shaped cross-sections and frequently contain lakes — like Haweswater in Cumbria.

The ice spread out from the dispersal centres of the Scottish mountains, Snowdonia-Plynlimmon, the Lakes, Donegal and Kerry, and slowly moving ice-sheets extended over a good deal of lowland Britain north of the Thames, depositing glacial till. This glacially transported material is completely unsorted, the particles ranging in size from those of finely ground rock-flour to rock masses larger than a house. It covers the Chalk of much of East Anglia, making up the monotonous gently undulating plateau of inland Norfolk and Suffolk. In the Aire and Ribble Valleys in Yorkshire, in the Firth-Clyde lowlands and in County Down, Northern Ireland, the till is sculptured into elongated, rather egg-like mounds, typically just less than one kilometre in length, known as drumlins. In Strangford Lough, south of Belfast, the drumlins have been partially submerged by a rise in sea level and they thus appear as tiny oval islands.

Erratics, boulders that have been carried some distance from their source, can often give a good indication of the direction of ice movement. Fragments of granite from Ailsa Craig, an islet in the Firth of Clyde, have been found along much of the east coast of Ireland, in the Isle of Man, Lancashire and Wales. Igneous rocks from Scotland and Norway occur in till in East Anglia.

The whole picture, however, is a complex one, and there have been at least three major advances of the glaciers in Britain over the last two or three million years —the Anglian, Wolstonian and Devensian (the most recent) glaciations. The ice-sheets moved in different directions and so the erratics found in the deposits associated with each advance are frequently quite characteristic. Between the cold periods there were warm 'interglacials' lasting several tens of thousands of years.

When the glaciers melted as the temperature rose at the end of each glacial episode the torrents flowing from the ice masses sometimes cut steep-sided gorges or meltwater channels. They are often now dry and have been compared to railway-cuttings. Examples include Newtondale in the North York Moors and the Loughaveema valley in County Antrim, Northern Ireland.

During the last glaciation much of Central and Southern England was free from ice; glaciers just entered East Anglia, for example, pushing up a ridge of tills or 'moraine' inland from Cromer in north Norfolk. The climate over the unglaciated portion of England must, however, have been very cold, with a mean annual temperature of perhaps minus 8°C (c. 18°F). The effects of these 'periglacial' conditions were considerable; permafrost — permanently frozen ground — existed, and one of the possible results of this has already been briefly noted. A characteristic feature of regions such as northern Canada and Spitzbergen that experience permafrost conditions today is the ice wedge. Vertical cracks in the surface layer fill with moisture in the brief arctic summer and this freezes when cold conditions return. The freezing causes expansion and thus the crack is enlarged, and so on. Some wedges are several metres in depth. When the ice melts, loose material is washed or blown into the cavity. Fossil ice wedges are clearly visible in many gravel pits in eastern England.

4. FOSSILS

Fossils and evolution

A comparison of the various theories concerning the origin of life is outside the scope of this book. Suffice it to say that it seems possible, perhaps partly through the production of substances containing carbon and nitrogen by certain types of volcanic activity, or perhaps as the result of electrical discharges (lightning) in the primitive atmosphere, that very simple life-forms may have been produced from non-biological sources.

It is now widely believed that the tremendous diversity of animals and plants that exists and has existed on the earth is derived from a few such simple forms through the process of evolution, the gradual change in the characteristics of organisms with the passage of many generations. Charles Darwin, in 1859, suggested that 'natural selection' was the principal mechanism of evolutionary change. Far more organisms are produced than can possibly survive to produce their own offspring. Within any population there is considerable variation; some plants and animals possess characteristics that give them a slightly higher chance of survival — the 'survival of the fittest'. As these characteristics are inherited over time unfavourable characteristics will be eliminated and combinations that are best adapted to the local environment will be reinforced. As conditions change, so populations change too.

Natural selection, however, is probably too simple an explanation to account for the great variety of living and fossil organisms; probably many other factors are involved. Nevertheless much of Darwin's work is fundamental to modern ideas on evolution.

Although evidence for evolution now comes from many branches of science, that of the fossil record is probably the most powerful. (Some of the other uses of fossils to the geologist have been mentioned on pages 20-23.) Fossils may consist simply of the hard parts of organisms, shells or bones, for example, that do not decompose readily and which were buried in a protective medium when the creature died. Sometimes, however, the remains were later altered: on the writer's desk is a fragment of 'petrified wood' from Idaho, in the U.S.A.; the structure of the plant tissue can clearly be seen but the original woody material has been entirely replaced by silica. Occasionally just the cavity formerly occupied by the fossil remains — a natural cast — the creature itself having been dissolved away. Very infrequently some of the soft tissues as well as the hard parts are preserved; such are the cases of insects and centipedes that have been found embedded in amber (fossil tree resin) or the mammoths preserved by natural refrigeration in the near-frozen swamps of Siberia. An almost entire specimen was found in 1907 and can now be seen in a Leningrad museum.

Fossils are rare in Precambrian rocks. Quite abruptly with the opening of the Cambrian a little less than 600 million years ago large numbers of quite advanced animals appear in the fossil records. This sudden burst of evolution was perhaps due to the proportion of oxygen in the atmosphere, which had been increasing throughout geological time, reaching a certain level, the 'first critical level' (about 1 per cent of the present quantity) so that more advanced forms were able to flourish in a way that had not been possible previously. Since then, although many forms and indeed whole groups of animals and plants have become extinct, the general trend has been for life-forms to become more complex.

A note on scientific names: scientific or 'Latin' names have already been met with (e.g. *Calamites, Lingula*). Every animal and plant, living or fossil, is given a name, usually derived from Latin or Greek, so that scientists of different nationalities can recognise what is being discussed. Very few fossils have common or English names and so scientific names will usually be used for fossils in the pages that follow.

Some common fossil groups

Corals have a relatively simple structure; the rock-building forms of the Carboniferous consisted of a cylindrical or conical skeleton of calcium carbonate surrounding an internal cavity. Many living and fossil forms are colonial, building reefs such as the Great Barrier Reef off the Pacific coast of Australia, and ultimately contribute to the formation of limestones. The corals appear first in the fossil records in the Ordovician and are extremely important for the zonation of the Upper Palaeozoic; a typical Carboniferous example is *Zaphrentis* (figure 10G).

Brachiopods form a small group today—there are less than 260 living species — but were much more important in the past for some thousands of fossil types have been described. They have a shell composed of two valves and a pedicle or stalk emerges either between these or through an aperture in one of the valves, fixing the creature to the sea-bed. *Lingula* burrows in the sands on the margins of warm seas and has remained unchanged since the Cambrian. Other forms like *Productus,* which has thick calcium carbonate shells, in some Carboniferous species up to 10 centimetres across, are abundant over a more limited range. *Terebratula* is a common Mesozoic brachiopod (figure 10E).

Even the simplest molluscs have a more sophisticated bodily structure than the brachiopods which some of them, the lamellibranchs, superficially resemble. The lamellibranchs or bivalve molluscs are abundant along Britain's coasts today; cockles, mussels and oysters belong to this group. They differ from the brachiopods in that a muscular foot emerges from between the two valves — some forms are highly mobile — and they have a more complex blood and nervous system. Sometimes the valves have an elaborate set of teeth and sockets, by means of which the two halves of the shell articulate with one another. One such shell commonly found in the Crag deposits of East Anglia is *Glycimeris* (figure 10C). Gastropods, the group that includes the snails, have only one, usually spirally coiled, shell. Both the

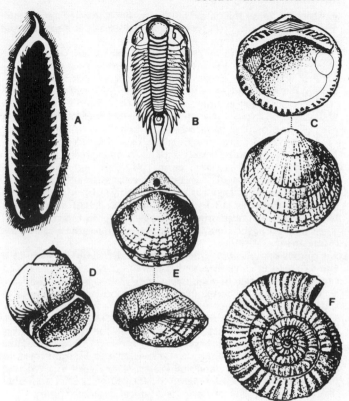

Fig. 10. Some invertebrate fossils: A **Didymograptus,** *Ordovician graptolite (x2); B* **Paradoxides,** *Cambrian trilobite (x1); C* **Glycimeris,** *Tertiary lamellibranch (x⅝); D* **Globularia,** *Tertiary gastropod (x1); E* **Terebratula,** *Jurassic brachiopod (x1); F* **Dactylioceras,** *Jurassic ammonite (x1); G* **Zaphrentis,** *Carboniferous coral (x½).*

45

gastropods and the lamellibranchs have existed since the Palaeozoic, but the two groups expanded greatly in numbers and variety during the Tertiary.

Ammonites are also molluscs, but their shells are coiled in one plane (plate 14) and are divided into chambers; as the creature grew it secreted larger and larger chambers, successively abandoning those that it had outgrown. The empty cavities were filled with gas and acted as buoyancy tanks. The ammonites were very abundant and evolved extremely rapidly in the Mesozoic and are used as zone fossils to subdivide the Jurassic. *Dactylioceras* (figure 10F) is a typical ammonite from the Lias. The group is extinct.

Trilobites are another group with no living representatives; they are arthropods, the phylum or major group that contains the insects, spiders, woodlice (which they superficially resemble), crabs and lobsters. They varied in size from a few millimetres to over 45 centimetres (18 inches) in length and had a segmented body and a broad head-shield, probably living on the sea-bottom. They are used in the zonation of the lower Palaeozoic and became extinct in the Permian.

Crinoids or 'sea-lilies' (they are not really plants) possessed a cup-like body or calyx surrounded by food-gathering tentacles and supported on a stalk of round columnar plates. These columnals, a centimetre or so across, are quite common in limestones of Carboniferous age.

Graptolites were small colonial animals; their colonies leave markings on fine-grained rocks rather like the blades of knives with serrated edges a few centimetres in length. They are quite abundant in certain Cambrian, Ordovician and Silurian rocks in Scotland, northern England and Wales. A few types survived into the Carboniferous. *Didymograptus*, the 'tuning-fork' graptolite that appears in the Ordovician, is a typical example (figure 10A).

The evolution of the vertebrates

Vertebrates are animals with backbones. The remains of creatures probably ancestral to the later vertebrates have been found in Ordovician rocks, but as a group they only became common in the upper Silurian and Devonian. Fossils of primitive fish-like creatures such as *Cephalaspis* (figure 11A), which had no true internal skeleton but had an elaborate external armour-plating, are quite common in the ORS rocks of Scotland. Scales and fairly complete specimens of much more highly evolved fish are known from the Permian Marl Slate from County Durham and the resistant teeth of sharks are quite common in the clays of the Mesozoic.

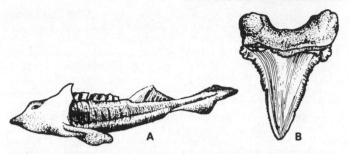

Fig. 11. Fossil fish: A **Cephalaspis** *(ORS); B the tooth of* **Carcharodon** *(Eocene shark).*

Some authors believe that the rising level of oxygen in the atmosphere cut off damaging radiation from outer space when it reached a 'second critical level' of about 10 per cent of the present amount late in the Palaeozoic. Until then, these writers maintain, life on the continents was impossible and was confined to the şeas. In any event it was in the Devonian that a second major burst of evolution took place and animals and plants began to invade the land masses on a substantial scale.

Amphibians, animals that require water for breeding but which are able to spend some of their life-cycle on land, appeared in the Devonian and became quite important in the Carboniferous. The reptiles developed in the Carboniferous, becoming the dominant vertebrate life-form in the Mesozoic. They had an important advantage over the amphibians in that they were able to breed without water, the young animals being confined for a time within the watery environment of an egg. Many different modes of life were represented amongst the Mesozoic reptiles: *Ichthyosaurus* and *Plesiosaurus* were marine and had a streamlined form for swimming. *Brontosaurus* and *Diplodocus* were probably land-dwellers and vegetarians, reaching an enormous size — 30 metres (*c.* 100 feet) in length and weighing up to twenty-five tons. *Tyrannosaurus* stood 10 metres high, had savage teeth 13 centimetres (5in) long and was carnivorous. Pterodactyls were flying reptiles. Footprints of some of these large reptiles are not uncommon, indeed a creature called *Cheirotherium* is known from its foot-prints alone — no bones have ever been found.

Both the birds and mammals arose from reptilian stock, but the early stages in the evolution of the birds are unknown. A creature called *Archaeopterix* has been found in Jurassic rocks in Ger-'

47

many; it had feathers but also reptilian teeth. Mammals, which produce their young alive, are warm-blooded and possess a sophisticated nervous system, first appear in the British fossil record in the Trias with a small animal called *Microlestes* but are of little consequence until after the extinction of the giant reptiles at the end of the Cretaceous. In the Tertiary they evolved as explosively as the reptiles had done in the Mesozoic, with the appearance of forms as varied as the whales, bears, sloths, elephants, horses, bats, and primates (monkeys, apes and men). Teeth of the mammoth, a large elephant that lived in Europe in the Pleistocene, have been dredged from the Thames at Westminster and the remains of hyaenas, bears and hippopotami have been taken from cave deposits in Yorkshire and the west of England.

The evolution of man

When Darwin wrote his book *The Descent of Man* in 1871 he advanced the theory that man had evolved from an ape-like ancestor, though there was at the time very little fossil evidence to support his thesis. The situation is now very different.

The relatively unspecialised 'lower' primates — tree shrews, lemurs and tarsiers — have been found as fossils in Eocene deposits in Europe and America. One species has been found in Hampshire. It is, however, from Africa that the fossils that tell of the more recent evolution of the human line have come. Forms such as *Aegyptopithecus* from the Egyptian Oligocene and *Proconsul*, a small creature from the Miocene of East Africa, are probably quite close to the direct human lineage. *Australopithecus* skulls have been found both in South Africa, in limestone caves at Taung and Sterkfontein, and in Kenya and show a large number of man-like features; they are serious contenders for the title 'missing link' between man and his ape-like ancestors. The australopithecines had a brain capacity of less than half that of modern man and probably did not make tools. All are several million years old. Some people have said, rather cynically, that man may have inherited his aggressive tendencies from *Australopithecus*, as many of the skulls are shattered and show evidence of violent death. Other anthropologists would take the view that this was not very likely, as in 1975 and 1976 it was revealed, on the basis of excavations in Ethiopia, that something very similar to modern man had co-existed in Africa with *Australopithecus* for two to three million years.

A number of different types of men appear to have developed in the Pleistocene, and opinions differ as to their relationship. The remains of Neanderthal man were first found in Germany but subsequently specimens have come to light in many places in

Britain — much evidence on their way of life has come from caves at Creswell in Derbyshire — as well as elsewhere in Europe and in western Asia and Africa. *Homo neanderthalis* seems to have been short and stocky with a backward sloping forehead. His brain was at least as large as that of modern man (*Homo sapiens*) and he made quite sophisticated stone tools, used fire and appears to have had a burial ritual. Some have maintained that Neanderthal man was ancestral to modern man, others that he represents a quite separate branch of the evolutionary tree. Hybridisation may conceivably have occurred between the two species. The Neanderthals seem to disappear quite suddenly from the record about 40,000 years ago, leaving *H. sapiens* as the only modern hominid type.

Evolving plant life

Some very ancient rocks contain simple plant fossils; in Rhodesia there are Precambrian limestones containing algae and a number of primitive forms have been found in rocks about 1,900 million years old in southern Ontario. It was not, however, until about the Devonian that vascular plants, plants with well organised structures and containing systems for transporting materials from one part of the plant to another, appeared on the scene. Before then the landscape of the continents must have been bleak in the extreme, devoid of life and resembling vast desert tracts or the 'rock barrens' of northern Canada.

Fossils of *Psilophyton* (ORS, Scotland) consist of a simple root-like structure from which spore-bearing stems extended upwards for a metre or so. By the Carboniferous there were forests of Lycopods, plants such as *Lepidodendron* (plate 15) with diamond-shaped scales on its trunk. Horsetails are today plants less than a

Fig. 12. A fossil fern (Carboniferous).

metre in height, with a central, often grooved and segmented stem and whorls of slender leaves radiating from each node. They are quite common in damp woods. In the Carboniferous forms such as *Equisetites* were very similar in structure but were 15 metres in height. Ferns such as *Neuropteris* are also common.

Cycads, with broad palm-like leaves spreading from a central trunk, are amongst the commonest fossil plants in Mesozoic rocks; good specimens can be obtained from some layers of shales and clay in the cliffs along the Yorkshire and Dorset coasts. The flowering plants, or angiosperms, which make up by far the largest proportion of non-microscopic plants today, evolved in the Cretaceous. Many trees such as the oak have remained almost unchanged since then, and the grasses and mangroves only became widespread in the Tertiary.

In the Ice Age there were great changes in the climate and major movements in the world vegetation belts took place. It is, however, often possible to reconstruct the vegetation of a locality for the various stages by the study of plant remains, particularly the tiny pollen grains, preserved in peaty deposits. Thus it can be shown that southern and eastern England was covered by a tundra vegetation of dwarf shrubs during the late glacial phase. As the climate became warmer this was replaced by a birch-pine assemblage and then by forests of oak, elm and alder. The clearance of the forest by early man can be documented in a similar way.

5. APPLIED GEOLOGY

Sources of energy

Man requires energy to power his means of transport, drive the machines in his factories, to pump water and to provide lighting as well as to heat his buildings. A very limited amount of energy is obtained from working animals, directly from the sun using solar cells and water heaters, the wind (windmills) and the tides (tidal power stations like the one on the Rance in Normandy), but most of man's power requirements are met by the combustion of 'fossil fuels'.

The energy released when coal, oil or natural gas is burned is energy that was derived from sunlight millions of years ago. Plants, both small marine forms and the large tree-ferns, photosynthesised in the Carboniferous as they do today, using the sun's energy to build up complex organic chemicals from carbon-dioxide and water. When the plants, and perhaps also the animals that fed on them, died and were buried in environments with little oxygen, decomposition was retarded; this was the first stage in the formation of our fuel supplies.

Coals originally formed as the result of the compression of peat beneath great thicknesses of strata. Brown coals or lignites represent an early stage in coal formation; they have a higher proportion of carbon than peat, as some of the 'volatile' elements (oxygen, hydrogen) have been driven off, but retain some moisture. Some of the structures of the original plant material remain. Lignite is only worth mining where 'higher rank' coals are not available; in Bohemia (Czechoslovakia) it is mined by opencast methods and is used to provide fuel for both power stations and railways. Bituminous coal is of higher rank, having undergone more extensive alteration; the proportion of water and volatiles is much lower. Most of the house and coking coals mined in Britain belong to this group. Anthracite is the end member of the series; it has undergone extreme compression and is usually about 95 per cent carbon. It is hard and burns slowly with great heat. Anthracite is mined in South Wales.

As explained in chapter 3 the coalfields of Britain have been extensively folded and fractured; coal-seams are dislocated by faults and may peter out into shales or divide. Sometimes the Coal Measures are concealed beneath hundreds of metres of more recent strata — the Kent coalfield, for instance, is not exposed at the surface at all, but is entirely covered by Jurassic and Cretaceous strata. The task of the geologist and his colleague, the mining engineer, is to determine the position of seams underground from surface evidence and from boreholes and nearby workings and to ensure that mining proceeds at minimum cost and in safety.

Petroleum was probably formed from the partial decay of animal and plant remains buried with sediments on sea or lake floors. The fluid oil, unlike coal, can migrate from its position of formation, accumulating in porous rocks such as sandstones and limestones in oil traps, where further movement is blocked by impermeable strata. An anticline provides a typical example of such a trap (figure 13). Oil has been obtained since about 1940 from sandstones in domed Carboniferous rocks beneath Permo-Trias cover at Eakring in Nottinghamshire. Natural gas forms and accumulates in more or less the same way as oil but is composed chiefly of lighter hydrocarbons such as methane (CH_4). Following the discovery of substantial quantities of gas at Lockton in north Yorkshire and in the Netherlands, drilling started in the North Sea in 1965. Large finds of gas associated with domes in Permian and Triassic sandstones beneath the sea followed and by 1970 many of the homes and factories of England had been converted to

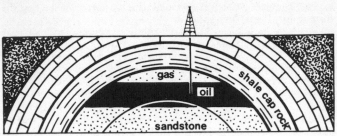

Fig. 13. Oil and gas trapped in an anticline.

North Sea gas. The main productive areas — namely the West Sole, Indefatigable, Hewett and Leman gas fields — are on the continental shelf off the coast of East Anglia, Lincolnshire and Yorkshire. In the early 1970s substantial oilfields were discovered in the rocks beneath the North Sea. By the end of 1976 seven UK oilfields were 'on stream': Argyll (the southernmost), Auk, Montrose, Forties, Piper, Beryl and Brent (north-east of the Shetlands). Exploration for oil and gas is continuing in the Celtic (Irish) Sea.

Nuclear energy contributes an increasing proportion of the world's electrical power. Fuel for atomic reactors like those at the Trawsfynydd nuclear power station in North Wales comes from radioactive minerals such as pitchblende, an oxide of uranium (UO_2), a black mineral that is sought for using a Geiger-Muller counter, an instrument that measures radioactivity.

There are important uranium mines in northern Canada, in Queensland (Mary Kathleen) and in Southwest Africa. Another area, significant both because it was the source of material on which Madame Curie did her early work on radioactivity and because it was an important source of uranium for the Soviet Union in the 1950s and 1960s, is Jachymov in western Czechoslovakia. Other uranium sources await development in several parts of Africa, in Australia's Northern Territory (Ranger) and in Western Australia (Yeelerie). Uranium has also been worked commercially on a small scale at the Trenwith and South Terras mines in Cornwall.

Geologists have been involved not only in the search for uranium, but also in the location of stable sites suitable for the disposal of radioactive wastes. Localities in the Scottish mountains and the remote deserts of Western Australia have been suggested. However, nuclear technology has undergone a substantial reassessment in the last few years. Serious doubts have been expressed about the wisdom of mankind becoming dependent for his energy supplies upon a technology that some consider exceedingly dangerous.

Ore minerals

An ore is a mineral or rock from which one or more metals may be extracted with profit or in hope of profit.

Two important ores of iron are the following:

HAEMATITE (Fe_2O_3) is a red mineral with a somewhat metallic lustre sometimes occurring in striking kidney-shaped lumps.

MAGNETITE, another oxide of iron (Fe_3O_4), gets its name from its magnetic properties which are usually sufficient to distinguish it from other minerals; it is black in colour and has a hardness of 6.

Haematite occurs in irregular masses or 'sops' in limestone in Cumbria; the limestone is dissolved by percolating water and replaced by iron oxide from the red Triassic rocks above. It also is the important constituent of the much leaner Jurassic sedimentary ores of the English Midlands.

Magnetite ores, on the other hand, are usually associated with igneous activity. One of the most famous magnetite ore bodies is that at Gälivare-Malmberget in northern Sweden, which has a mining area some 6.5 kilometres long and an average iron content of 60 per cent.

Minerals containing copper, tin, lead and zinc are frequently found close together; some of the ores are listed below:

CHALCOPYRITE : this is a sulphide of copper and iron ($CuFeS_2$); it has a rich golden colour but often shows an iridescent tarnish. It is rather soft (hardness: 3.5-4); when crystals occur they are tetrahedral in form.

ZINC BLENDE (ZnS) is red-brown in colour, brittle and has a resinous lustre. It has a toffee-like appearance.

GALENA (PbS):this is an important ore of lead. It is very heavy, having a specific gravity of 7.5, and has a bright metallic lustre when freshly broken but soon tarnishes to a dull lead-grey. Crystals are cubic.

CASSITERITE (or tinstone) (SnO_2): this is red-brown to black in colour and has a brilliant lustre when fresh. The hardness is about 6, specific gravity 7. In Cornwall it sometimes appears in the form of needle-like crystals.

These ores occur in veins (infilled fissures) emanating from igneous intrusions such as Dartmoor, the Cornish granites and igneous masses beneath the Pennines, along with 'gangue' (pronounced 'gang') minerals like fluorite (CaF_2), barytes

(BaSO4), calcite and quartz. The veins are intruded into the country rock as fluids and different minerals crystallise at different temperatures—ores of tin at over 550°C, copper at about 500°C, those of lead and zinc at rather lower temperatures. In any vein, therefore, the different ores are encountered at different depths. Thus in the Dolcoath Mine near Camborne, Cornwall, copper was worked close to the surface, below was a narrow zone where both copper and tin were encountered and below this, closest to the igneous source, a region in which tin was found alone. Sometimes the hot fluids in the veins reacted with limestone through which the vein passed, and the ore body now spreads out laterally as a 'flat'. In parts of Weardale, for example, some of the Yoredale limestones are impregnated with ore for some distance on either side of the main veins.

Following the erosion of some of the ore-bearing veins, some of the crystals of cassiterite were often transported by streams to lower ground in the form of rolled pebbles. The working of valley gravels for tin began in prehistoric times in Cornwall and the scars and sinuous ridges of waste left by those who turned over the gravels searching for stream tin on Dartmoor in nearby Devon from about 1150 onwards are a familiar feature of that heather-clad upland.

The small but often very rich veins of base-metal ores are now largely worked out (plate 13) and in Britain and elsewhere attention is now being concentrated on much larger but lower grade ore-bodies. At Palabora in South Africa, for example, working an ore body with an average copper content of 0.57 per cent is profitable. Another large copper deposit at Coed-y-Brenin in North Wales was described in April 1973 by the company that had been investigating it as containing 0.3 per cent copper. In this case, partly because of the very low values and partly because of resistance to opencast mining in the Snowdonia National Park, it was decided that its development would not be profitable.

Gold (Au) is very inert and is found in the metallic state, sometimes alloyed with silver. It is a rich yellow colour, very heavy (specific gravity 19.3) and soft (hardness 2.5-3.0). It may occur in substantial masses (nuggets) or in fine grains dispersed in rocks.

Gold, like other metal ores, occurs in veins associated with igneous intrusions and has been mined in Britain and Ireland since prehistoric and Roman times. Most British gold contains a fair amount of silver. Several mines in Cornwall (e.g. Wheal Sparnon Mine, Redruth) and Cumbria have yielded small quantities of gold, but by far the richest gold area in Britain is the Mawddach River district of North Wales, where a number of mines worked from 1840 until the end of the last century. The

Clogau (St David's) mine provided gold for recent royal wedding rings and the fleur-de-lis for the Prince of Wales's new crown.

Gold-bearing grains, like fragments of tin ore, are sometimes redeposited in streams to form 'placer' or 'alluvial' gold. Appreciable quantities of alluvial gold exist in deposits beneath the silt of the Mawddach estuary, and it was suggested that these might be workable, but damage to the beautiful landscapes of the Snowdonia National Park would have been so enormous that the proposal to dredge for them was abandoned. Relatively small amounts of placer gold have been obtained at Leadhills in southern Scotland, from Sutherland and Dartmoor. The conglomeratic 'banket' deposits of the Rand in South Africa are really a very ancient (and enormous) placer deposit.

Sometimes appreciable quantities of gold are extracted on the refining of other ores; thus the ore mined on Bourgainville island in Papua New Guinea contains 0.7% copper and about 1g per tonne of gold and 2g per tonne of silver.

Water supply

Water is often obtained from wells drilled into aquifers (water-bearing strata) such as the Chalk and Lower Greensand in eastern and southern England and some of the Triassic sandstones in the north. The Thames Valley is in fact an artesian basin, the water accumulating in the porous Chalk capped by London Clay and underlain by Gault Clay. Originally the water came to the surface under its own pressure, but it now has to be pumped. Many aquifers have been depleted by over-pumping and increased emphasis is now being placed on surface waters and above-ground reservoirs. But the siting of the reservoirs and their dams — the selection of water-retentive and stable sites — constitutes another task for geologists.

Building materials

A very wide range of rocks has been used as building stones in Britain. Granite has traditionally been used in Aberdeenshire and parts of the west of England; Carboniferous sandstones are met with in the walls of many a Yorkshire cottage; many of the Oxford colleges are built from local Jurassic limestones. Quarrying of slates for roofing has long been traditional in North Wales and has continued at Delabole, St Teath, Cornwall, since the time of Elizabeth I.

The Oxford Clay, particularly between Bedford and Peterborough, is the most suitable raw material for brick-making in Britain, as small quantities of bituminous material present in the clay effect a saving of fuel during firing and other impurities present assist with the fluxing process.

Cement is made by firing a mixture of crushed Chalk or limestone and clay; several cement works are sited close to the base of the Chalk escarpments in eastern England as the lower layers of the Chalk contain considerable amounts of clay and the rock may virtually be placed in the kilns as it is. The expanding use of concrete in construction has increased the demand for sand and gravel as well as cement; substantial quantities are obtained from the Thames Valley and the valleys of East Anglia where abandoned, water-filled pits are a common landscape feature.

Environmental conflicts

As Britain's population grows and living standards rise, demand for cement, bricks, sand and gravel for the building of homes increases; with industrial expansion requirements for fuel, metals and other raw materials like salt, potash, anhydrite and china clay similarly grow. Tip-heaps and opencast workings are unsightly and might be regarded as particularly out of place in National Parks such as the Lake District, Snowdonia, the North York Moors or Dartmoor. Where processing of raw materials takes place at the point of production problems of air and water pollution arise (for example the milky coloration of many rivers and streams in south Devon and near St Austell in Cornwall is the result of contamination by the china clay undertakings (plate 12)). Yet most of these upland areas of Britain are regions of high unemployment where traditional activities like hill farming are declining and there are those who see a large investment by a mining company in terms of more employment, a boost for local trade and improvements in local services. Also, the development of British mineral resources can achieve substantial savings in imports and bring about an improvement in the country's balance of payments position. The decisions before County Planning Officers and the central Government in cases of natural resource development are often difficult.

Sometimes land can be re-used after the minerals have been removed. When the National Coal Board adopted opencast methods for winning coal in south Northumberland the topsoil, subsoil and rocky overburden were piled separately. After the coal was removed they were replaced carefully and the area was under cultivation once again within two years. Abandoned clay and gravel workings near Cambridge are used for water sports and angling; similar flooded pits in Kent and Bedfordshire are now managed as nature reserves and wildfowl refuges.

6. DISCOVERING GEOLOGY FOR YOURSELF

The equipment is simple: boots and warm clothing, some newspapers for wrapping specimens and a notebook and pencil are all that are essential. Many geologists have a special hammer with a square cross-section and a pick at one end, such as that in plate 4. A hand-lens is helpful.

Rock and fossil specimens may be collected from road-cuttings, quarries, contractors' temporary excavations as well as natural exposures of rock such as sea-cliffs and the sides of river valleys. Be considerate when you visit inland, well-weathered exposures at beauty spots visited by members of the public; it takes years for the natural weathered surface to develop and a few thoughtless hammer-blows can be very disfiguring. There has been unfavourable press-comment following the battering of Haytor Rocks on Dartmoor by large numbers of untrained 'geologists'! Obviously this does not apply to quarries where the face is still being worked or sea-cliffs where erosion by the waves is constantly exposing fresh material.

If you visit an active quarry it is absolutely essential that you obtain permission. Apart from anything else the stone is usually won by blasting with dynamite. This will mean that parts of the

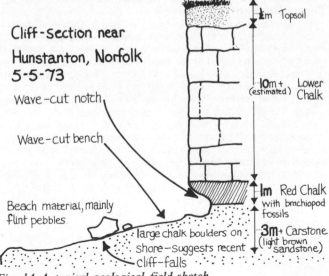

Fig. 14. A typical geological field-sketch.

quarry are very dangerous at certain times of the day or on certain days of the week. There may also be other dangers such as unstable faces, trackways, machinery and so on of which you could have no knowledge. If permission is given you may be asked to sign an indemnity form; by completing this you are accepting responsibility for your personal safety while you are on the company's premises. If permission is refused (which is not very often) you can be sure there are good reasons.

When collecting don't despise fragments at the foot of the face; exposure often makes fossils 'weather out' and they show up clearly. Try to split sedimentary rocks along the bedding planes; you are more likely to find fossils this way than by indiscriminate bashing. It is sometimes worth taking a large stone containing a fossil home for careful chipping rather than attempt to extract the fossil in the field.

Old mine tips are almost always worth looking over. Good crystals of fluorite, quartz and calcite as well as specimens of ore minerals like galena and zinc blende are not difficult to find on many tips in the Welsh Borderland, the Derbyshire Peak District, Weardale, Devon and Cornwall. Fossil plants and lamellibranchs are sometimes to be found on the heaps near coal mines, though permission is again essential for work on many of these as they may still be used for dumping and some heaps have smouldering fires inside them.

Follow streams in upland regions — you are likely to come on steep places where bare rock is exposed and fossils can be collected easily. Note the direction and amount of the dip of the rocks, if any, and whether the rocks change their character as you proceed. As one walks along the upper Tees for example the different beds of the Yoredale Series — limestone, shale, sandstone — alternate in the bed of the river and the point where it crosses the Whin Sill dolerite is quite obvious.

Specimens should be labelled with the exact locality — 'Limestone band 1.5 metres above stream, 50 metres north of Styles Bridge, Gurgleswick, Yorks'. Or you can mark the specimen with a number and record the details separately in a notebook. Take notes also about the locality, recording details of the rocks present, their appearance and thickness, any fossils found. Figure 14 illustrates this procedure.

Examine, sketch and photograph landscapes. When out in the hills, for example, ask yourself questions such as 'Has this valley been glaciated?'; 'Why do so few streams cross that hillside?'; 'How did that lake form?'

APPENDIX

Chemical formulae

The chemical elements, of which about one hundred are known, are the basic building blocks of matter; the smallest particle of an element is an atom. Elements combine with each other in definite proportions to form compounds. Although a few elements occur naturally in the uncombined state, for example gold, sulphur, copper and diamond (pure carbon), most minerals are compounds of two or more elements. They are often represented by chemical formulae, a sort of international shorthand for chemical structure; each element has a symbol, and small figures below the symbols give the ratio of the different elements in any one molecule (compound atom) of the substance. Thus the mineral fluorite—a mineral occurring in veins with metal ores—is written CaF_2; this means that for every atom of calcium there are two of the element fluorine. The symbols for a number of elements are given below:

Aluminium	Al	Manganese	Mn
Barium	Ba	Nitrogen	N
Calcium	Ca	Oxygen	O
Carbon	C	Phosphorus	P
Chlorine	Cl	Potassium	K
Copper	Cu	Silicon	Si
Fluorine	F	Silver	Ag
Gold	Au	Sodium	Na
Hydrogen	H	Sulphur	S
Iodine	I	Thorium	Th
Iron	Fe	Tin	Sn
Lead	Pb	Uranium	U
Magnesium	Mg	Zinc	Zn

GLOSSARY

Anticline: arch-like upfold in rocks, in which strata dip outwards from an axial crest (see figure 3 and plate 1).

Aquifer: water-bearing rock.

Basalt: dark coloured, basic volcanic rock.

Batholith: large, dome-like, often elongate, intrusive mass, usually of granite.

Bedding: the arrangement of sedimentary rocks in layers.

Cleavage: the splitting of crystals and some rocks (e.g. shale) along planes of natural fissure.

Corrasion: the wearing away of the sides and bottom of a stream by the mechanical action of transported material.

Cross-bedding: small-scale layering in sedimentary rocks in which the individual layers are at an angle to the main bedding planes (plate 5).

Crust: outermost layer of the earth; the portion that is accessible to direct investigation.

Crystal: solid bounded by smooth plane surfaces, the positions of which are determined by the internal arrangement of the atoms.

Cyclothem: stratigraphical unit consisting of a series of beds of differing character that are repeated in an almost regular way, as in the Yoredale Series or Coal Measures.

Dyke: mass of igneous rock filling a vertical fissure and cutting across bedding planes in surrounding strata (plate 8).

Erosion: the gradual wearing away of the land by weathering (q.v.) and the removal of the material by rivers, glaciers, etc.

Erratic: mass of rock differing in character from the surrounding strata, transported by ice to its present location.

Escarpment: an inland cliff or steep slope, usually formed by the erosion of inclined strata or faulting (figure 7 and plate 10).

Evolution: irreversible modification of organisms with descent.

Fault: fracture in the earth's crust along which there has been movement.

Fossils: remains or traces of past life, preserved by natural processes in the rocks of the earth's crust.

Geosyncline: major subsiding depression in the earth's crust in which depositional material accumulates, eventually forming great thicknesses of sedimentary rocks.

Glaciation: the covering or partial covering of land by glaciers or ice-masses.

Granite: coarsely crystalline, acid, plutonic rock, often comprising batholiths. Although many granites are clearly igneous, having solidified from the molten state, it is argued that some granites originated by the replacement of existing rocks and might be referred to as metamorphics.

Igneous: rocks that have solidified from molten magma.

Joint: crack in a mass of rock along a plane of weakness along which there has been little or no movement, c.f. Fault.

Laminae: thin, plate-like layers.

Mantle:layer between the earth's crust and core.

Metamorphism: alteration of rocks by heat or pressure.

Meteorite: mass of rock or metallic material from outer space.

Mineral: natural inorganic substance with a definite chemical composition.

Moraine: (1) material transported by a glacier; (2) landform, often ridge-like, formed by deposition of such material.

Ore: mineral or rock from which a metal can be economically extracted.

Periglacial: zone marginal to a present-day or former glacier or ice sheet.

Plutonic rocks: those that solidified slowly at a great depth in the earth's crust.

Rift valley: downfaulted area between two parallel faults.

Schist: rock formed by intense metamorphism, containing mica and displaying good cleavage.

Sedimentary: rocks composed of sediment — mechanical, chemical or organic; secondary rocks formed as the result of weathering and the re-deposition of material from pre-existing crystalline rocks.

Sill: igneous mass intruded along a bedding plane in a pre-existing rock.

Slate: dark coloured metamorphic rocks showing good cleavage.

Stalactite: icicle-like structure, usually composed of calcium carbonate, formed by the evaporation of percolating water dripping from the roof of a cave.

Stalagmite: upward-pointing structure of calcium carbonate built up on the floor of a cavern on the evaporation of water dripping from above.

Syncline: trough-like downfold in rock strata.

Tor: isolated rock mass, usually but not necessarily formed by the weathering of granitic rocks (plate 6).

Volcanic rocks (or extrusive rocks): those igneous rocks that have cooled rapidly at the earth's surface and are therefore fine-grained, e.g. basalt.

Weathering: the chemical decomposition and mechanical disintegration of the rocks *in situ*, i.e. without appreciable lateral movement.

FURTHER READING

The following paperbacks will take the reader a little further:
Britain's Structure and Scenery; L. D. Stamp; Collins (Fontana), 1960.
Geology and Scenery in England and Wales; A. E. Trueman; Penguin Books, revised edition 1971.
Fossils; F. T. Rhodes, H. S. Zim and P. R. Shaffer; Hamlyn, 1965.
Rocks and Minerals; H. S. Zim and P. R. Shaffer; Hamlyn, 1965.
The next three volumes represent a slightly more formal approach to the subject:
A New Geology; M. J. Bradshaw; English Universities Press, 2nd edition 1972.
Essentials of Modern Geology; M. Smith; Phillips, 1968.
Beginning Geology; H. H. Read and J. Watson; Macmillan, 1974.
These books are more specialised and will lead the student a good deal further into the subject:
The Geological History of the British Isles; G. M. Bennison and A. E. Wright; Arnold, 1975.
The Principles of Physical Geology; A. Holmes; Nelson, 2nd edition 1972.
Understanding the Earth; I. G. Gass, P. J. Smith and R. C. Wilson (editors); Open University and Artemis, 2nd edition 1974.
The next volume will prove most interesting to those interested in the problems of land-use conflict and the social and economic factors involved in the development of mineral resources:
The Politics of Physical Resources; P. J. Smith; Open University and Penguin Books, 1975.

Also useful are the publications of the British Museum (Natural History), entitled *British Palaeozoic Fossils, British Mesozoic Fossils* and *British Cainozoic Fossils,* which are beautifully illustrated and should allow almost any fossil likely to be found in the British Isles to be identified. The same museum produced a handy and well illustrated series of leaflets on specialist topics: *British and Irish Gold* would be useful to anyone planning a little amateur gold-prospecting as a holiday pastime, while *Fossils in Caves* would be of value to someone interested in caverns.

The Institute of Geological Sciences publishes *Regional Memoirs* for each part of the country: *East Anglia and Adjoining Areas, South-West England, The Southern Uplands* and so on; the appropriate memoir would provide an excellent introduction to the geology of your home area. The Institute also publishes geological maps of much of the country. Sometimes both 'Drift' editions, showing deposits such as glacial till and valley gravels, and 'Solid' editions, from which these superficial deposits are omitted, are available.

INDEX

Printed by C. I. Thomas & Sons (Haverfordwest) Ltd., Press Buildings,
Merlin's Bridge, Haverfordwest, Pembrokeshire.